BEEKEEPING BUZZ

The Beginning Beekeepers Guide to Their First Hive

Nicole Wrinn

ISBN: 1497591430
ISBN-13: 9781497591431

CONTENTS

Reality Check

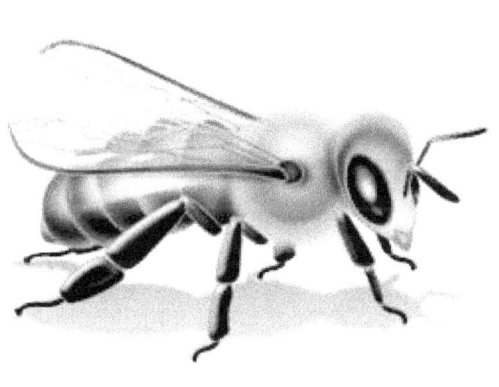

The prospect of keeping bees as a business or a hobby can be thrilling, but before you go out and order a colony, you need to consider a few things. First of all, why do you want to go into beekeeping? Many people see selling honey as a lucrative enterprise, but they don't realize how much time and money they'll have to invest in it. And frankly, to be a good beekeeper, you need to like bees.

To avoid future financial and emotional disappointment, we recommend starting small. Keep bees as a hobby at first. Build, say, two or three hives, and treat the activity as an experiment. Familiarize yourself with the equipment, handling techniques, and honey processing methods. Moreover, get used to being around a large number of buzzing, stinger-equipped bees.

If there are any experienced beekeepers in your area, you might want to speak with them about what it takes to be successful in this venture.

Secondly, take inventory of your yard/garden. Do you have a suitable location for at least two hives? How much space do you have for hives, flowering plants, and honey processing?

Lastly, are any of your family members or neighbors allergic to bee stings, or would they be uncomfortable with a lot of bees flying around? Will your outdoor pets be safe around so many bees? Some districts even have laws or ordinances prohibiting beekeeping, so make sure you act within the law.

Also, are you prepared to be stung numerous times? Are you willing to care for the bees as you would any other animal? They are prone to disease, so you must keep the hive(s) clean on a regular basis. You also be able to troubleshoot any problems. You can't just let them leave them alone for days on end and expect quality honey. They need maintenance—not daily, but regularly—provided by a keen eye and a gentle hand.

So, if you think keeping bees will be safe, interesting, and financially feasible for you, it's time to get started. We hope this little book will help you understand the basic concepts of beekeeping, enabling you to raise your bees in a clean, nourishing environment. Remember, happy bees make the sweetest honey!

Chapter 1: Apiology 101

The first step towards becoming a beekeeper lies in understanding the biology and behavior of honeybees. Lacking this basic knowledge, you will be endangering yourself as well as the people around you, not to mention the bees.

A beekeeper must act as a "bee manager." That is, he/she should continually supervise the bees and apply the scientific knowledge necessary to anticipate bee behavior. Over the course of several seasons, this knowledge will develop into an intuition: you will simply know when changes are going to occur and what to do if problems arise.

Bee Society

As you probably know, honeybees live in colonies, each headed by a queen. Since colony management forms such a crucial part of beekeeping, it's advisable to learn how colonies operate.

A honeybee COLONY resembles an agrarian caste system in human society, consisting of a few to tens of thousands of workers (sexually immature females), a few to several hundred drones (sexually mature males), and a single queen (a sexually mature female). A combination of fertilization (or lack thereof) and nutrition determines a bee's caste.

Though they cannot reproduce and are the smallest in size, WORKERS perform numerous tasks in the hive. Depending on their age, they first serve as housekeepers, cleaning and polishing the cells and removing debris; then as nursemaids, feeding the brood and caring for the queen; then as construction workers, building beeswax combs and sealing gaps with wax and propolis (resin mixed with wax), or as maintenance workers, ventilating the hive; then as grocers, accepting incoming nectar; then as undertakers, removing the dead; and then as guards, willing to sacrifice their lives in defense of the colony (they die upon injecting their barbed stingers). Their final task (when they're 3 weeks old) involves foraging for pollen, nectar, water, and propolis. If reared in the spring, they will live and work for about six weeks during summer before they die; if in the autumn, they will remain with the colony for several months.

During late spring or early summer, DRONES form a sort of male harem to mate with a new queen during her mating flight, providing sperm to fertilize her future eggs. As with workers that use their stingers in defense, the mated drones die instantly. The largest bees in the colony, the unmated drones stay on until autumn, when they're banished from the hive due to their inability to work and their drain on the food supply. Workers normally feed them, but they can feed themselves—only in the hive, however. Once forced out, they will starve.

The QUEEN has one function: to reproduce and expand the colony. She can be distinguished by her elongated abdomen, relatively short wings, and long stinger. Within about 48 hours after mating, she begins her egg-laying career, which will last about two to three years. As mentioned above, workers build beeswax combs. The individual cells vary in size. If a worker places the Queen's egg in, for instance, a small, worker-sized cell, she will release some stored sperm to fertilize it. If the worker places it in a large, drone-sized cell, she will not release sperm, leaving it unfertilized.

Workers feed the larvae discriminately too, giving future workers and drones mostly honey and pollen, while giving potential queens (if necessary) almost an entire diet of "royal jelly."

The term BROOD refers collectively to all three developmental stages: egg, larva, and pupa. The Queen usually lays one EGG per cell. A worker affixes it to the bottom of the cell. At first the egg looks like a grain of rice, standing straight up. The egg then starts to bend over and on the third day hatches into the white grub known as a LARVA. Workers feed the larvae for five to six days, depending on the caste, and at the end of that period they cap the cells with brown, convex discs of beeswax. Inside the capped cells, each larva stretches itself out and starts spinning a cocoon, becoming a PUPA. During the pupal stage, the baby bees develop into their adult forms. Workers and drones emerge after 12-14 days, while queens come out after only a week. This cycle continues for as long as the colony exists and can be an indicator of the colony's health.

Beeswax combs are not only used to rear the brood, but also to store "bee food": honey (containing carbohydrates) and pollen (containing protein). How do the bees construct combs? Well, the workers secrete wax from their abdominal glands; the wax hardens into scales on their abdomen. Using spines on their legs, the bees remove the scales, manipulate them with their mouth, and add the wax to the cell imprints on the foundation (see Chapter 3).

Seasonal Changes

Weather and seasonal changes provide helpful indicators of bee behavior. In temperate climates, the bees' "New Year" begins around September, when the flow of nectar and pollen slows down significantly. At this time, the colony gradually ceases to expand and actually starts to diminish, as the old bees die off without being replaced by new generations. The queen stops laying eggs, the construction workers seal up with hive with propolis, and the bees banish the drones (see above). Once the temperature falls to 57° F, the bees cluster around the brood, keeping it warm (at about 93° F). The cluster expands with a rise and contracts with a drop in temperature. The bees on the outside of the cluster help insulate the bees on the inside, which feed on the stores of honey and pollen. When the weather warms up, the cluster can move over the comb to reach more honey. If the cold lasts too long, however, they will consume the current store and not be able to reach the other food, and will starve.

A well-stocked colony will be able to feed the queen while she's in the cluster. Then she can resume laying eggs in late December or early January, and the dead bees can be replaced. Such a colony will emerge in spring strong and healthy. If a colony does not have sufficient stores, its population will be much reduced by spring, and the queen won't be able to lay more eggs until the workers gather new pollen and nectar when the flow begins again.

Upon the arrival of spring, the bees will become quite active. In addition to foraging for pollen and nectar, they will also collect water to cool the hive and liquefy honey to feed the brood.

As spring progresses, the population quickly grows, and foragers will be out gathering nectar and pollen, often in surplus. When the temperature rises high enough, the workers will start building drone cells. They will also select a few larvae to become queens (see below). Eventually the hive won't be large enough to hold all the bees, and some will start clustering outside the entrance. In spite of overcrowding, the colony will attempt to build more storage combs as long as there is space.

This is also the season for swarming. That is, before any of the new queens emerge and the nectar flow reaches its peak (between March and June), the current queen and about half of the colony will fly out of the hive during the warmest part of the day and cluster on a tree limb or other object. The bees send out scouts to find a location suitable for a new hive. After an hour or so of reconnaissance, the scouts report back and lead the rest to the site. The bees then build combs and gather nectar and pollen, while the queen lays eggs.

Back at the parent colony, the work goes on as before: gathering supplies, caring for the brood, guarding the entrance, and building combs. When the first queen emerges, she eats honey, grooms herself, and then seeks out the other queens. They engage in a fight to the death. The champion remains in the hive for another week before flying out to mate with several drones.

As in spring, the bees gather water to cool the hive. They must keep the temperature at about 93° F at all times. They spread water inside the cluster so that it will evaporate.

Once the colony reaches its peak population, the focus shifts to preparing for winter. The long days allow the bees to collect surplus amounts of nectar and pollen, increasing the honey stores. As a beekeeper you must be careful to leave enough honey for the bees to consume during the winter.

The Queen's Role

Normally, when you acquire a colony of bees, it will already possess a queen, usually marked with a small dab of paint. In some cases you may need to introduce a new queen or you will observe the colony preparing for a new queen itself. Therefore it's important that you understand the dynamics of queenship in honeybee colonies.

As the sole provider of new workers, the queen is responsible for the strength of the colony. If you discover that a colony is functioning poorly, look to the queen first before considering other possibilities, such as disease. In a healthy brood, the queen will lay eggs consistently, missing only a few cells. The caps will not have punctures. In all, there should be four times as many pupae as eggs, and twice as many pupae as larvae. This ensures that the workers will develop properly and at a manageable rate in proportion to the food supply.

The egg-laying season occurs in spring and early summer, when the queen may lay as many as 1500 eggs per day. Egg production slows down towards the beginning of autumn and won't resume until the following spring.

But the queen does more than just lay eggs. The colony bears her chemical signature. If she produces a large amount of the pheromone, "queen substance," the colony will respect her leadership and will work together as a team. If she and her mates possess beneficial genetic qualities, her offspring will also have those qualities.

What happens, then, when the queen fails in these tasks?

Emergency and Supersedure

Let's begin our discussion with a colony in need of a new queen due to death, loss, or removal. In this case, the workers will have to raise an "emergency queen" from young (less than 3 days old) worker larvae in a modified, vertically-hanging cell. Unfortunately, in an emergency situation the workers don't have time to feed the new queen the usual amount of royal jelly, so she may not possess the best qualities as an adult.

If the current queen can no longer perform her central functions in the hive, the workers prepare to supersede her. They select fertilized eggs and modify the cells to hang vertically. When the eggs hatch, they will feed the larvae royal jelly almost exclusively. Sometimes the new queen will emerge while the old queen is still living, but the latter will no longer be in charge of the colony.

The new queen leaves on her mating flight about a week after emerging from the cell. She travels a fair distance from the hive. Drones find her by following the pheromones she releases. She mates in-flight with seven to fifteen of them before returning to her colony. As we discussed in Chapter 1, she begins laying eggs about 48 hours later. While she lays the eggs, workers feed her royal jelly (a milky substance containing digested honey and nectar or pollen, mixed with a fertility protein) and take care of any other needs she may have.

NICOLE WRINN

Chapter 2: Location, Location, Location

When setting up your hives, you need to take into account several factors with regard to location and environment.

Ideally your bees should be able to find water within a quarter of a mile and nectar and pollen within a mile of the hive. They can fly farther to reach those necessities, but traveling bees may cause problems for your family and neighbors, who will surely complain about the numerous droppings on their cars, the groups of bees visiting (and sometimes drowning in) their pools or birdbaths, and the increased bee activity in the area.

To avoid this kind of drama, exercise good colony management techniques. Buy or rear a strain of bees with a gentle temperament—those that don't sting too readily or hover too long. Take note of your bees' flight pattern and, if necessary, modify it by planting a hedge or fence that is at least six feet tall. This will prevent them from flying too close to people and vehicles. You might also want to surround your apiary with a fence or hedge to keep out prying eyes. If you have a suitable roof, such as that over an apartment building, consider establishing your apiary up there. In addition, have a shallow container of water ready and waiting for your bees when they emerge from the hive in the spring. If you don't provide this necessity immediately, they'll find water elsewhere, and good luck trying to get them to switch back! Keep an ample water supply available from March through October. Avoid locating your apiary in areas where the hives and/or forage will be endangered by pesticides.

On the flip side, too much human and animal traffic can bother your bees, so it's advisable to keep them in a secluded area. Avoid assembling a hive with a top entrance, as the removal of the supers (hive bodies with honey) will result in numerous disoriented bees. Although swarms pose little threat to humans, keep this natural behavior as invisible to your neighbors as possible. See Chapter 6 for more handling techniques.

If your apiary stands twenty feet or closer to a property line, place some kind of barrier (e.g. fence, hedge, etc.) between the hives and the boundary; it should be at least five feet high. If your apiary stands thirty feet or closer to a sidewalk or road, place a similar barrier between the hives and the public area or elevate the hives so the bees will fly above people and vehicles.

As a beginner, you probably won't have more than two or three hives, but as your apiary grows, do not exceed four hives per quarter of an acre. If you want to set up your apiary in an "outyard" (i.e. a place not on your property), select a site that isn't already populated with other beekeepers' hives.

Make sure you can easily access the hives year-round, especially when it's time to harvest the honey. You don't want to wear yourself out by having to wheel a bunch of honey up a hill or drive several miles to the field where your apiary is located.

As for the hives themselves, you don't want them too high (wind) or too low (damp). Windy conditions will dry out and disturb the bees, while constant moisture increases the likelihood of disease. The proper height will vary depending on where you live. Don't set the hives on the ground, but raise them to allow air to flow underneath, keeping them dry. If mice, skunks, and raccoons are a problem, elevate your hives to about 16 to 18 inches above the ground and/or add an entrance reducer (see Chapter 3). The bees can defend themselves, but it's best to keep their life as stress-free as possible (it's already hard enough making honey!). Another advantage from a taller height is that the bees won't sense vibrations as keenly. Alternately, you can leave the hives 4 to 6 inches off the ground, as long as the bees won't suffer from cool temperatures and high humidity. The main considerations here are bee safety and hive accessibility (particularly when the hive bodies are laden with honey).

Situate your hives in location where the bees will receive partial sunlight, such as beside a tree. Don't place the bees in deep shade or direct sunlight. Try to provide natural protection against the summer and winter winds (hills, buildings, tall trees). Make sure each hive is on dry, level ground, with the bottom boards sloping slightly for rainwater to drain out. Throw some mulch around the base of the hive to discourage weeds and grass from growing over it.

Lastly, turn each hive so that the entrance faces the southeast, where the sun rises every morning. This will ensure that the bees get to work early each day.

Chapter 3: Suit 'Em Up!

Now that you've found a good location for your apiary, you need to buy your equipment and purchase or build your hives. We recommend that beginners ask an experienced local beekeeper for help in assembling the hives. They should also obtain all their equipment before acquiring bees. We strongly advise against novices buying used equipment, which may have unconventional dimensions, contain pathogens, and be of indeterminate value. Whether you buy new or used equipment, always make sure it has been inspected by your state's Apiary Inspection.

The Hive

Unless you're naturally talented with building, you'll probably want to purchase an assembled or unassembled hive. If the hive requires assembly, make sure you follow all the manufacturer's directions. You may need to buy some additional supplies, like nails, for assembly.

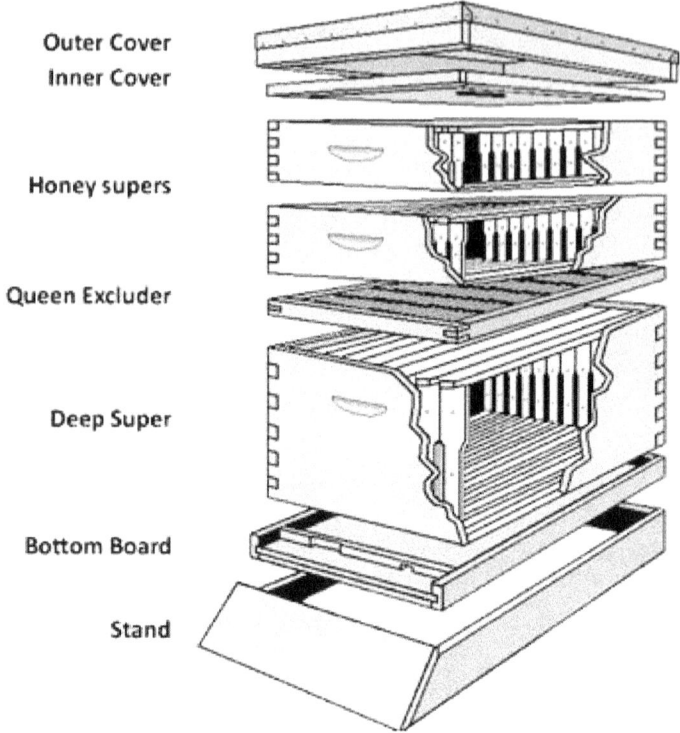

Outer Cover
Inner Cover

Honey supers

Queen Excluder

Deep Super

Bottom Board

Stand

Let's review the basic components of a typical manmade beehive, from bottom to top. First, there's the hive stand. Not all beekeepers use a stand, and it can vary in height. You can construct it from bricks, concrete blocks, logs, railroad ties, or pallets, or you can simply buy a commercially produced stand. As we discussed in "Location, Location, Location," the stand serves to elevate the hive from the ground to increase air flow, discourage undergrowth, and protect the bees from honey-hungry predators.

Next you have the bottom board, the floor of the hive. Remember, this should slope slightly for drainage to keep the hive dry. The bottom board has an opening in the front through which the bees will enter. If you trouble with predators, you can add an "entrance reducer" to narrow the opening.

Above the bottom board sits the deepest hive body. It holds 10 frames and can be of various depths. Though they can be used to store honey, the deepest bodies are most suitable for containing the brood nest. Most beekeepers use either two full-depth bodies or one deep and one shallow. If you are installing a package of bees (see "Bee Mine"), you may want to start with a single full-depth body. It is possible to buy hive bodies that hold only 8 frames.

Some beekeepers will insert a "queen excluder" between the brood nest and the honey supers (hive bodies). Usually a perforated sheet of metal or plastic, or a wire grill, it acts as a filter, allowing only worker-sized bees to pass through. It's not a necessary piece of equipment, but if you're having troubling finding the queen and/or want to separate the honey combs from the brood combs, you might consider adding one. However, don't insert the excluder until after the bees have started storing nectar in the supers.

As you know by now, each frame holds beeswax combs used either for storage or brood rearing. The frame itself is made of wood or plastic. As a foundation for the combs, you fit into the frame a sheet of beeswax or plastic imprinted with a cell pattern. Beginners should start with thick, heavy foundations for the frames in the brood nest. For the honey supers, use thin foundations in shallow bodies, as these are best for producing cut-comb honey (the easiest for a novice). Fasten the foundation at the top with the wedge of a "wedge-type" top bar or with a bead of wax to a grooved top bar.

Choosing a frame and foundation combo can be difficult. Should you go with plastic beeswax foundation in plastic frames or pure beeswax foundation in plastic or wooden frames? While plastic has gained popularity among beekeepers, you need to consider how much you can spend, how much assembly will be required, how long you plan to use the items, and whether you're willing to replace non-durable items. Just keep in mind that using beeswax foundation in wooden frames will necessitate reinforcing the structure. As one option, you can to insert metal hooks or pins in the top bar and run vertical wire down into the bottom bar, keeping the foundation centered. As another, you can electrically embed horizontal wires into the foundation. But no matter what combination you decide on, you need to check that the foundations won't become dislodged in their frames. In doing so, you'll ensure that the bees will draw the comb properly.

The inner cover rests atop the uppermost super. Not only does it prevent the bees from gluing down the outer cover with propolis and wax, it insulates the hive, protecting it from harsh summer rays and icy accumulation. Many beekeepers add a "bee escape" to the center hole to help remove bees from the supers.

Lastly, there is the metal-plated outer cover, which fits over the inner cover and shields the hive from all kinds of weather. It also aids in stacking hives one on top of another.

We don't advise buying plastic hive components (other than frames and foundation), as these don't breathe as well as wood and can become warped.

The bees will coat the inside of the hive with propolis, but you'll need to paint the outside yourself. Use an exterior, light-colored paint (latex or oil-based). If you have multiple hives, you may want to vary the colors to help the bees' flight orientation.

Smoker

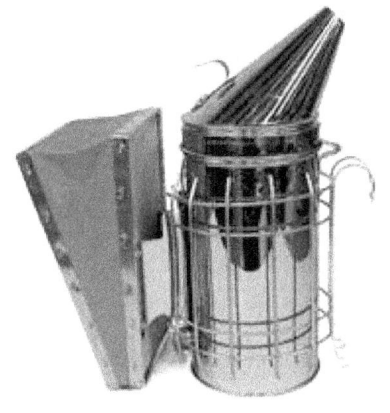

We'll go deeper into the uses of the smoker in "Colony Management," but for now you should familiarize yourself with its components. All smokers consist of a metal fire pot, a grate, and bellows. Most beekeepers use a 4 x 7 inch model with a heat shield around the firebox and/or a hook for hanging it over the hive while they work.

To operate the smoker, place a small amount of fuel (wood shavings, pine needles, dry leaves, corn cobs, twine, rags, burlap, etc.) on the coals above the grate, light it, and squeeze the bellows until a flame appears. Add more fuel while you squeeze the bellows; stop when you've filled the fire pot. See that the smoke is cool enough, and if necessary add a little damp grass or leafy material to keep the fire merely smoldering. Watch out so your smoker doesn't throw sparks or flames near the bees, as these can singe their wings and body hairs! Refill the smoker as needed once the smoke grows weak or hot, or turns blue. Pack the fuel with your hive tool.

The coolness of the smoke calms the bees, masks alarm chemicals, and triggers a feeding response where the bees start eating honey, thinking the hive might be on fire. A bee can't sting when her belly is full!

Hive Tool

A hive tool enables you to separate frames and hive bodies, as well as scrape away wax and propolis. When you're working the bees, carry it in your hand at all times and use it gently so as not to frighten the bees. Also, so be sure to keep it clean to prevent the spread of disease: periodically stab it into the dirt or burn it in the smoker.

Protective Clothing

Your beekeeping hood (or hat) and veil are essential for reducing stings. If you choose a hat, make sure it has a wide brim so that the veil will hang down away from your face. Depending on your preference, you may combine these with coveralls or a windbreaker/pullover. Wear light-colored clothing made of smooth, tear-resistant fabric, along with boots or shoes with light-colored socks. Bees don't like dark colors or rough fabric. Secure your ankles by tying up your pants or tucking them into your socks, and protect your wrists by tying up your sleeves. The fabric of your clothing should be breathable in hot conditions.

Although it's tempting to wear gloves to prevent stings on your hands, they will only impede your movements, agitating the bees unnecessarily. However, if at first you're uncomfortable using your bare hands, try form-fitting gloves, as these will give you a lighter touch while keeping your hands free of stings and honey/propolis residue. Avoid wearing any cologne, aftershave, hair spray, etc. when handling the bees, as they may react adversely to the scent. Launder your beekeeping outfit regularly to get rid of any odors that might upset the bees.

Cost

The cost of your setup will vary depending on how many hives you want, whether you'll assemble or build them yourself, and whether you buy your tools and clothing individually or in kits. Below you'll find a list of potential equipment, along with the general price range. The items in red are mandatory for safe, successful beekeeping.

Item	Price
Complete Beekeeping Suit (Hood, Veil, Coveralls, w/ Gloves)	$40-$300
Beekeeping Jacket (Pullover, Hood, Veil, w/ or w/o Gloves)	$30-$50
Beekeeping Gloves	$10-$20
Hive Tool	$5-$15
Smoker	$15-$50
Queen Excluder	$15-$450
Beginner's or Starter's Kit (Hive, Smoker, Hive Tool, Hood, Veil, Gloves	$150-$500
Assembled Hive Body (8-10 Frames)	$50-$500
Unassembled Hive Body (8-10 Frames)	$15-$100
Beekeeping Tool Kit (Smoker, Hive Tool, Uncapping Tool, Beehive Brush, Scraper, etc.)	$40-$60
Frames	buy in bulk
Foundation	buy in bulk

Chapter 4: Bee Mine

All right, you've got all your equipment set up. What kind of bee should you get? "Honeybees, of course! Isn't that obvious?" Not exactly. Honeybees come in many varieties, so choosing the right kind for you has a huge impact on the success of your venture. Let's review the characteristics of the major races available for purchase:

	Italians	Carniolans	Caucasians	Russians
Scientific Name	*Apis mellifera ligustica*	*Apis mellifera carnica*	*Apis mellifera caucasica*	none (hybrid)
Color	Light yellow or brown with brown/black abdominal bands	Grayish brown with brown spots or bands on abdomen	Dark to black with grayish abdominal bands	Darker than Italians, lighter than Carniolans
Availability	Easy to find	Not common	Not common	Rare
Temperament	Gentle	Gentle, quiet	Very gentle	Slightly aggressive
Swarming Tendency	Moderate	High	Moderate	High
Hygiene	Good	Moderate	Moderate	Good
Flight Orientation	Poor; tendency to drift	Good	Poor; tendency to drift	N/A
Productivity	Early brood rearing; good foragers; excellent honey comb	Early brood rearing; good foragers even on wet/cool days; need	Not good at producing honey but can forage at low temperature	Good foragers; good honey producers; need strong nectar flow

In addition to Russian, there are many other hybrid races, such as Midnite, Starline, Buckfast, Buckeye, West Virginia, and Minnesota Hybrid, bred specifically to exhibit certain desirable traits. Do your research to find the best breed of honeybee for your area, taking into consideration the climate and pollen/nectar availability. Also consider your preferences as a beekeeper and choose the race you believe will be easiest to handle. As always, consult fellow beekeepers, as well as association publications, but stay true to your budget and ability.

Package Bees

Once you've decided on a race of bees, you need to determine the method of acquiring them. Regardless of the method, you should have your bees delivered in early spring to give them a chance to establish themselves in your hive(s). And though you can buy a colony established by another beekeeper, beginners should start with either a package or a nucleus colony (nuc).

Packages provide the best opportunity for new beekeepers to learn how to handle bees, as they are cheap, good-natured, and not prone to brood diseases. You'll be ordering your package from a southern state or California. Packages usually weigh two to three pounds, with each pound representing about 3,500 adult bees. A newly mated queen will come with the package.

In order for the bees to arrive in time for spring, you should order them in January or February. You can install the package either on drawn comb (containing honey and pollen), in which case you should have the bees delivered in early April, or on foundation, in which case the bees should arrive in late April, early May. The main determinant here is the temperature: bees installed on foundation won't be warm enough yet to reach food in early April in northern areas.

You can obtain packages from local suppliers or by mail. If by mail, your package will be shipped in a large wooden cage ventilated with wire screen. During shipment, the bees feed on a solution of 50 percent sugar syrup contained in a can in the middle of the cage. The queen comes in a separate cage near the feeder can, along with a couple of attendants. She eats sugar candy.

Ask the post office to notify you immediately when your package arrives. Inspect the bees to make sure there are no more casualties than normal. If all is in order, you can begin preparations to install them. Keep the cage warm and shield it from the wind, but do not heat it directly. Cover the cage for transport. You need to install the bees within 48 hours of receiving them; in the meantime, feed them the 50 percent sugar syrup solution at room temperature, sprinkling it on the wire screen. Then store the bees at 50-60 °F (the temperature in your basement, porch, or garage).

Install the package in late afternoon or early evening. First, open the designated hive and remove about half of the middle frames (in a 10-frame body). Add an entrance reducer and a temporary plug made of green grass to the hive entrance.

The best method of installation is by shaking the bees into the hive. Before opening the package, sprinkle the wire screen all over with sugar syrup. Jar a corner of the package to knock the bees down into the bottom. Next, remove the square piece of wood covering the top of the package cage, and take out the feeder can. Sprinkle the bees again with sugar syrup. Inspect the queen and set her cage aside in partial sunlight. Then shake the bees from the package into the waiting hive; make sure you clear every corner of bees. The bees' wings should be damp from the syrup, so they won't fly away. Leave the package cage abutting the hive entrance overnight in case there are any stragglers. Use your hive tool to gently level out the layer of bees on the bottom board.

Now, take out the cork from the candy end of the queen's cage. Make a tiny hole through the queen's sugar candy: watch out, don't poke her! Turning the cage candy-end-up, wedge it between the top bars of two frames in the hive. The bees will then be able to attend to her and she'll be able to send chemical signals to them.

If you installed your bees on foundation, you'll need to feed them sugar syrup until the nectar flow becomes sufficient to sustain them. Even with nectar available, it generally takes four weeks for the population to reach self-reliance because initially more workers will die than emerge (new workers need 21 days to mature). You may expand the hive for more brood and/or honey after one and a half to two months.

During this settling-in period, you must provide adequate food and conduct regular inspections. To feed the bees, invert a feeder can or plastic jar with a perforated lid over the hole in the inner cover. Then cover this with an empty hive body and the outer cover. Use syrup medicated with fumagillin to combat nosema disease (see Chapter 5). Refill the feeder if necessary, but wait about a week before examining the hive. At this time, remove the queen's cage and make sure that she's laying eggs and that the colony has accepted her.

Nucleus Colonies

Nucs are small, easy-to-handle colonies (originally from an established colony), containing only four or five frames, a new queen, and a good stock of workers, brood, and food. They don't cost as much as an entire established colony, though they are more expensive than packages. You can inspect nucs before buying, but there is still the possibility of disease, so you should acquire the colonies only from reputable local beekeepers. It's also wise to have a beekeeping authority inspect the nucs after purchase. That said, starting with nucs can be worthwhile for a beginner. With good weather, strong nectar flow, and some supplemental feeding, you might even get a honey crop the first year. Before you know it, your once-empty hive will be teeming with bees!

Capturing Swarms

Another option, perhaps a little intimidating, is to capture a swarm of feral bees. You'll probably be getting over twice as many bees as in a purchased package, and they'll all be free. Plus, they're likely to have a wide gene pool and be well-adapted to your area. However, you need to be careful, not to mention well-prepared! We therefore advise beginners to ask an experienced beekeeper for assistance.

As you learned in Chapter 1, bees typically form swarms in springtime when the hive becomes overcrowded. The queen and about half the colony rush out of the hive and cluster on a nearby tree limb, bush, fence, or building. Protected by your beekeeping gear (see Chapter 3), you'll need to either cut down whatever the swarm is clinging to or shake or scrape the bees into a container: a hive or cardboard box works well, but make sure it has a tight-fitting lid! If the swarm is too high for you to reach safely or too large to control, let it go and find another one. Don't take unnecessary risks. Also, it's best to arrive on the scene when the swarm has just clustered, because if you wait too long the scouts will have found a new location and the bees will fly away.

When you've got the swarm contained, take it back to your apiary and empty the bees into a hive without frames. Capturing the queen along with the others will ensure that the swarm will make the hive their home. You should also give the bees some combs with honey, pollen, and/or brood to encourage them to establish a colony.

If you don't want to go out in search of swarms, you can get the swarms to come to you. Purchase some pheromone lures and place them in special bait hives or empty hives; you can add combs, but you don't have to. Set up the bait boxes out in the open, at least a story high. Check the boxes at least once a week.

So what's the catch? Well, "adopting" feral bees carries with it much the same uncertainty as adopting feral cats or dogs. They may not be resistant to disease; they may not be very gentle in temperament; and they'll probably have a tendency to swarm. Also, they may produce less honey. This is because "wild" bees haven't gone through selective breeding like the commercial races. Although well-suited to the local climate, they will likely exhibit some unwanted behavior. You'll have to observe them for a time in order to determine how best to handle them. Nevertheless, when you become more experienced in beekeeping, you have the option of breeding them to bring out a particular feature, such as hardiness.

Chapter 5: BeeMD

At the same as you order your package or nucleus colony in January/February, you need to formulate a plan to manage potential pests during the upcoming season. If you did your homework, your bees should arrive with a clean bill of health, but the possibility remains that they may contract a disease or be endangered by pests on your property. Therefore you should learn how to detect early signs of a problem and quickly implement a remedy.

Brood Diseases

American foulbrood (AFB) strikes larvae fewer than 53 hours old. They ingest the bacterial spores with their food and die after their cells are capped. Because the spores get in the food, this disease will spread through the hive like wildfire. And when this weakness becomes apparent, robber bees will steal the contaminated honey and infect their own colony. If when examining the larvae you see a "pepperbox" pattern of dark larvae and brown cells with punctured caps, you'll have to get a prescription from a veterinarian for Tylan®, and if that doesn't work, you'll have to burn the entire hive, including the bees. That's why you've got to prevent the spread of AFB at all costs. You can try applying Terramycin® (pre-mixed or self-mix) to your colony in the fall after you've removed the honey supers and in spring at least 45 days before you put them on again, but this drug is not guaranteed effective.

Note: Stop feeding antibiotics to your bees at least six weeks before any surplus honey flow to ensure that the honey you harvest is entirely free of drugs.

European foulbrood (EFB) kills larvae between two to four days old, before their cells are capped. A "stress disease," EFB comes into play at the peak of brood rearing in spring/early summer. It can disappear or hang around for a considerable time. EFB will not destroy the colony but will greatly reduce its population. Look for larvae that have turned from white to yellow and appear twisted in the cell. You can try to chase away the disease by stimulating a nectar flow or requeening the hive with a young queen. Alternately, you can treat EFB with Terramycin® (pre-mixed or self-mix). However, under the influence of the medication, some infected larvae may survive and continue to contaminate the hive.

Sacbrood occurs as a virus during the first half of the brood-rearing season. It doesn't present a serious threat to the colony as long as the dead larvae are removed (either by the bees or by you). Signs of sacbrood are larvae with upright, dark heads, and failed pupae that look like fluid-filled bags. These will be scattered among healthy cells. Nurse bees and robber bees can transmit the disease. A strong nectar flow will usually clear things up, but you can also try replacing combs or requeening the hive.

The fungal disease chalkbrood turns larvae into chalk-white, cottony mummies in late spring, near the height of brood-rearing. High temperatures usually drive this disease out of the colony, and nurse and worker bees will remove the dead. Unfortunately, chalkbrood can hang around in your hive equipment for years; the bees will pass the fungal spores to each other quite easily. It won't destroy the colony, but it will severely stunt its growth. If you've seen signs in a hive, change out the combs and/or equipment. You may also want to replace the queen. You can try to prevent chalkbrood by keeping the hive warm and dry; also, increase drainage and plug up any holes.

Adult Diseases

The single-celled organism called Nosema can be quite deadly. The adult bees ingest its spores through food or water, and they become weak with dysentery. Infected workers defecate all over the hive and can't produce as much brood food, and an infected queen can't lay as many eggs. The colony may attempt to replace her. And the worst part is, you won't be able to detect the presence of the disease until infection is widespread.

As your newly installed package will be highly susceptible to nosema, you should feed the bees sugar syrup medicated with Fumidil-B® (fumagillin) at least 30 days before the honey flow, but stop administering the drug at least four weeks before the honey flow.

Adult bees can also develop the chronic bee paralysis virus (CBPV) and the acute bee paralysis virus (ABPV). Be careful not to mistake infected bees for robber bees, as they will be attacked by guard bees near the hive. Yet unlike robber bees, they won't fight back. Paralytic bees tremble constantly and can't fly. Their body hair falls out, giving them a dark, shiny, or greasy appearance. They will try to climb up surfaces but will tumble back down.

There's not much you can do to treat paralysis, but don't worry, it won't kill your colony, even if it does diminish production. The colony will recover eventually, sometimes after a short period. If these viruses keep cropping up, you can requeen the colony and/or replace old combs.

Dysentery can result from too much water in a bee's diet, the presence of nosema, a damp hive, or winter confinement. To prevent dysentery, don't give your bees food with a high water content; don't keep them cooped up too long during winter but allow them to defecate outside; and keep the hive ventilated.

Parasitic Mites

Varroa mites will give you no end of trouble, so you should do all you can to protect your apiary against them. These external parasites feed on the blood of bees in all stages of development, and they can destroy your colony if not dealt with. On the plus side, you'll be able to see the mites. The females (reddish-brown, the size of a pinhead) attach themselves to adult bees and suck their blood. This parasitic feeding causes the bees difficulties in flying and working, and shortens their life span. The females later enter the cells of young larvae, hide in the food at the bottom, and wait until the cells are capped and the larvae have finished spinning cocoons. The mites feed on the larvae. Three days later, they lay several eggs. Then the females and offspring feed on the developing bee larvae. After about a week, the new females mate with the new males; the males die upon copulation. The females will then emerge from the cell with the bee (if it survives the pupal stage) and live in the hive for several months.

In examining your hives on a routine basis, look for a spotty brood pattern, malformed workers and drones (particularly in the wings), crawling bees, and discarded brood at the entrance. Pupae will also appear mottled.

How do you control varroa mites? Beekeepers have traditionally used pesticides, such as Apistan® and Checkmite+®, but these are risky and not always effective.

To exercise safe beekeeping practices, you should use pesticides only as a last resort. There are several ways to inspect a hive for mites, including using an ether-roll or powdered-sugar roll to separate mites from the bees. We prefer setting up a mite-trap on the bottom board of the hive. Place a piece of white cardboard on the bottom board and coat it with a sticky material. Take a piece of 8-mesh/inch hardware cloth and set it on top of the white board, stapling strips of cardboard around the edges so that it will sit about a quarter of an inch above white board. Leave the trap in the hive for three days. Then count the number of mites on the board and divide by three to determine how many mites fell to the board per day. If you have about 50 mites per day in late summer, you should go ahead and treat the colony. At all other times you should take into consideration other factors that might be contributing to the infestation.

You have several treatment options, starting with the introduction of a more mite-resistant stock of bees. In early spring, you can add frames of drone foundation to encourage the bees to produce drones. Once the cells are capped, remove the frames and freeze the developing brood for one to three days, killing the mites. Then you can throw the frame away or remove the dead drone pupae and use it again.

In addition, you can continually use the sticky boards (described above) to trap the mites and remove them, or you can replace the bottom board with a special screen so the mites simply fall out of the hive. However, you still have to worry about the mites within the capped cells.

Tracheal mites feed on the bees' blood and affect them in a similar manner. The colony can die from such an infestation and will be most vulnerable during the spring when there are mostly old workers. Unlike varroa mites, tracheal mites can't be seen easily with the naked eye. As small as specks of dust, they penetrate bees' tracheae and feed and reproduce inside them. Bees spread the mites to each other, and the colony's population shrinks. If you suspect a tracheal mite infestation, you'll need to send a sample of bees (stored in alcohol) to your local cooperative extension office.

You can treat your bees by using Mite-A-Thol® at a temperature above 70°F for 30 to 40 days. When the bees inhale the menthol vapor, the mites in their tracheae become dehydrated. Don't use this during nectar flow.

You can also make a four-inch grease patty using two parts sugar to one part Crisco. Place it between two sheets of wax paper, cutting off the excess paper. Then set it on top of the middle frames. When the bees get covered in the oil, the mites will become confused in selecting hosts. Since the oil is harmless, you can use grease patties for a long period of time.

Pests

If bacteria, viruses, and parasites weren't bad enough, you may have to deal with small hive beetles, ants, wax moths, and foraging mammals.

The small hive beetle emerged as a major bee pest in Florida in 1998. Currently it's prevalent only in the southeastern US, but if you order bees from one of those states, you could be getting beetles along with them. The beetles generally can't take over strong, healthy colonies. Nevertheless, your new package will be in a vulnerable state just after installation. The female beetles lay their eggs on or near combs; the larvae eat pollen, wax, bee eggs, and bee larvae. The adults will also eat the colony's honey stores and defecate in the honey, causing it to ferment and run out of the supers.

To crack down on these beetles, make sure you don't leave honey too long in storage. Also, try treating the soil around the hive (where the pupae form) with an insecticide or use Checkmite+® inside the hive. You can even move the hive to a location with a different type of soil (beetles prefer sandy soils).

Ants don't usually bother beehives, but if a group of hungry ants does set up shop inside, it will be difficult for you to evict them. You can keep the ants at bay simply by elevating the hive, removing any vegetation or debris beneath the bottom board, and adding fuel oil to the ground underneath.

Wax moth larvae like to chew on the wax, cocoons, discarded skins, and pollen of old combs. They bore through the cells, spinning a webby mass as they go. When they form their cocoons, they chew into the wood, weakening the frame. The best defense against wax moths is to store your honey in freezing temperatures (as low as 0°F). Not only will freezing kill all stages of wax moth, it will also prevent the honey from crystallizing.

In the fall, mice move into apiaries located in less-populated areas. They build their nests in the corners of the lower hive body, away from the bees' cluster. To make room, they chew combs and frames, often urinating on them. The bees don't suffer any physical harm, but they end up with unusable material in the spring.

You can keep mice from entering your hives by adding entrance reducers or 3 mesh/inch hardware cloth. If there are already mice in a hive, remove them and their nests and reduce the hive entrance. Don't wait for the bees to repair damaged combs or you'll get a bunch of drone cells; replace all frames and combs yourself, as needed. Cover up combs in storage with impenetrable materials.

In contrast to mice, skunks, opossums, and raccoons will actually devour your bees. They attack at night, every night, and as a result the colony remains on full alert. Additionally, they leave bee remains all over the place, scratch up the hive entrance, and tear up the ground surrounding the hive. Strong colonies will defend themselves fairly well, but weak ones won't survive. You can deter these predators with screens, queen excluders, entrance reducers, and elevation of the hive.

If you live in an area heavily populated by bears, you may need to take other precautions. Please go to the "Further Reading" section at the end of this book for more information.

Chapter 6: Handle with Care

Let's face it: in this business you're going to get stung. And those stings hurt! But if you want to separate yourself from the newbie who's all thumbs and welts, you should learn the essentials of safe beekeeping.

First of all, take a deep breath and realize that as long as you move carefully around the hives, the bees will leave you alone. They're busy with various tasks; you're just the supervisor checking on their progress and health. The bees know what they're doing, so you should act like you know what you're doing too. In other words, don't come barging in, but don't be too tentative either. Work quietly and confidently.

You need to examine your hives at the proper time of day. The bees will be most amenable on a warm, clear day between 10 A.M. and 4 P.M, when the older workers are out collecting nectar and pollen. Due to a strong nectar flow, the populations will be smallest in spring. Summer can be tricky, as the flow of nectar may become intermittent. Be careful during a lull in the flow, as bees from stronger colonies may attempt to rob the stores of weaker ones. A threatened colony will post more sentries to guard the entrance. Besides robbing periods, the bees will also be more aggressive due to the cooler temperatures on rainy days and in the early or late hours of the day. As the population grows towards its peak during summer, overcrowding often occurs. Also, the colony will have started building drone cells. Look for signs of an imminent swarm; you definitely don't want to get caught up in that! When the days grow long enough, the bees will be focusing on the coming winter and will spend a lot of time out foraging.

In Chapter 3 you learned how to light your smoker. Now we're going to explain how to use it. On days when most of the workers are out in the field, you won't need to apply much smoke. On cool, cloudy days, you'll probably have to use a fair quantity to soothe the grouchy bees.

Make sure all your movements are as gentle as possible; strong vibrations will scare the bees. Don't leave the hive open for more than ten or fifteen minutes. Make your examinations thorough but brief, and if the bees get upset, close the hive immediately. Also, honey combs will attract robbers from neighboring hives, so don't let them in or the hive may be compromised.

When approaching a hive, blow a couple of puffs into the entrance and into any other openings. Give the bees a few seconds to settle down before removing the outer cover. Don't start your inspection at the front but come around from the back or the side. Blow more smoke under the outer and inner covers as you lift them up. Lay the outer cover upside down on the ground (you can stack the honey supers on it later). If there are any bees clinging to the inner cover, knock them off at the entrance or just leave them if there aren't that many. You can place the inner cover on top of the boxes you've removed or simply set it aside.

After exposing the topmost honey super, blow a little smoke across the tops of the frames to drive the bees down into the hive. Give your smoker a puff each time you lift out a frame or set one back down. Starting with the second frame in on the side you're working from, use the straight end of your hive tool to break the propolis seal between the frames. Slowly lift the frame out, careful not to scrape the bees on the one next to it. After examining it, you can set the frame aside for a larger workspace only if there's no possibility of robbing; place it against the hive near the entrance. Pry the rest of the frames apart in the same manner: gently break the propolis seal with your hive tool, move the frame into the open space left by the first frame you removed, slowly lift it up, and then return it to its original position. Once you've put all the frames back, replace the first frame you removed.

Blow some more smoke as you remove the honey super(s) in order to inspect the brood chamber (usually the deepest hive body, located on the bottom of the hive). Repeat the process described above, but when you examine the second frame in this time, look for the queen. If she's on the frame, set the frame down against the hive, take the next one out, set it down, and then replace the queen's frame (don't crush her!). For the other frames in the brood chamber, take them out one at a time. If you want to inspect the brood itself, turn your back to the sun and tilt the frame so the sunlight falls gently on the comb. If you want to see the other side, turn the frame so the top bar is vertical, rotate the frame 180°, and then bring the top bar back into a horizontal position. Do this once more before putting the frame back. Once you're done examining the brood, replace the first frame you removed.

To reassemble the hive, repeat the steps above in reverse order. Make sure you puff your smoker and wait a few seconds before replacing each hive body and cover.

First Aid

"Okay, I did all that, but I still got stung!" We warned you, didn't we? In any case, don't panic and keep your clothes on—unless you want to get more stings!

A honeybee's stinger secretes a chemical alarm signal, attracting more bees, so leave the apiary as quickly as possible. Do NOT remove any clothing until you've exited the area.

You should recall from Chapter 1 that worker bees have barbed stingers. Because of this, you should NOT try to pull out a stinger: that will simply cause more venom to enter your body. Instead, scrape the stinger away with your fingernail or a pair of tweezers. Wash the sting site with water as an extra precaution.

Do NOT scratch the itchy area. To reduce the swelling, apply ice. If stung on your arm or leg, elevate it.

Pain, swelling, and itching are normal. Allergic reactions involve more severe symptoms, such as trouble breathing, dizziness, a swollen tongue, nausea and vomiting, or hives. If you experience any of these, call 911 at once. Otherwise, take acetaminophen or ibuprofen, or, if you're over eighteen, aspirin. You can also take an antihistamine or apply baking soda and water, or calamine lotion.

The symptoms should subside in four to five days. Sterilize the sting site regularly to prevent infection.

Multiple stings may indicate aggressiveness in your bees, but you can prevent the situation from escalating by smoking and closing the hive. If you receive a lot of stings, you will experience the same symptoms as if you were stung once.

Chapter 7: Beekeeping Yearbook

The last chapter provided some general guidelines for handling bees. Now we're going to discuss how to manage your colony through seasonal changes. We'll start with the period from December/January, as this is when you should order your first colony of bees as a beginning beekeeper. Once your colony has gone through a full season, you can begin implementing advanced colony management techniques. For now, let's stick with the basics. Don't tinker too much; give your bees a chance to do their thing.

Late Winter/Early Spring

After ordering your bees in December/January, figure out what diseases, parasites, and pests you should be concerned about for your particular situation. Lay out a plan of management; then order or build whatever supplies you need.

After installing your bees, don't open the hive until the temperature rises above 40°F, and then only for a few minutes and at noon on a sunny day. Also, don't disturb the bees' cluster. If it's still cold in March/April, feed your bees sugar candy or dry granulated sugar. (Next season you can feed them honey, but you don't have any as yet). When the temperature is above 50°F on a windless day, inspect the hive quickly for brood, disease, and food stores; then close it up again. The bees must stay warm to survive.

As long as the temperature remains above 40°F (i.e. the bees can break cluster), you should feed your newly installed package a sugar syrup solution consisting of two parts sugar and one part hot water until the nectar flow picks up. Once more forage is available, you can reduce the proportions to one part sugar and one part water. The sugar triggers the queen's egg-laying mechanism; the bees will begin to rear brood and expand the colony. However, the bees will become dependent on the syrup, so don't stop feeding it to them too soon.

To feed the bees syrup, you can invert a pail or jar over the top bars of the frames in the upper hive body, or over the hole in the inner cover. Perforate the lid of the feeder with a nail so that a vacuum will form inside. Make sure the syrup doesn't leak. Cover the feeder with an empty super, and then cover the hive itself. If it's windy out, weigh the cover down with a brick or other heavy object.

At the same time you're feeding the bees sugar syrup, you'll also need to feed them a pollen substitute to supply them with the nutrients they need for rearing brood. However, bees will only eat a pollen substitute if they're already bringing in some pollen and if you place it directly in their midst in the brood chamber. Otherwise, they won't touch it. Also, the substitute won't increase production, but it will prevent it from ceasing.

You can make a pollen substitute by mixing 1 oz. of brewer's yeast and 10 ½ oz. of sugar together with 5 ½ oz. of hot water. Then add 6 oz. of soybean flour until the mixture has the consistency of peanut butter. Check the label on the soy flour so it doesn't contain high sucrose or a lot of stachyose (a toxin for bees). Press the mixture between two sheets of wax paper to form a patty.

To give the substitute to your bees, remove the wax sheet from the top of the patty, turn the patty over, and place it over the cluster of bees on the top bars. Turn the inner cover upside down to make room for it. Check the patty every couple of days and replace it with another just before the bees have completely consumed it.

Continue to observe the colony through April. In four to six weeks after installation, you should see fully drawn combs and several frames of honey. If the queen is running out of room to lay eggs, you may need to add a second brood chamber.

Late Spring/Summer

At this time you should also be on the lookout for signs of swarming. Since your package came with a newly mated queen, you probably don't need to worry about supersedure yet (unless she's having problems), but if the colony is growing rapidly enough you may see preparations for a split. To prevent or at least reduce the potential for swarming, make sure the bees have plenty of room to rear brood and store nectar. Check the colony every eight to ten days for queen cells on or near the bottom bars of the brood frames. If you find any, destroy them and prepare to split the colony yourself.

"Split the colony? I'm not sure I could do that..." Yes, it does require some skill. However, there are many ways to keep the situation from getting to that point. In early spring, you'll probably find the queen in the uppermost hive body. When the workers start making honey, she may run out of room up there, triggering a swarm response ("We need more space!"). All you have to do is note when the brood nest is getting overcrowded, and then reverse it with an empty hive body. The queen can then work her way up without space constraints.

When the dandelions and maple trees bloom, add a super to the hive (if you haven't already). Separate the brood nest from the honey supers with a queen excluder. Don't remove honey supers until the combs are properly sealed.

During a strong nectar flow, always stay one step ahead of the bees: if 6-7 of the 10 frames are full and you see fresh white wax along the lower edge of the top bar, add another super immediately. The empty combs will stimulate foraging and honey production. You should add the new supers to the top of the hive (oversupering).

In adding new supers, keep in mind that bees would rather go ahead and store nectar and honey in drawn combs as opposed to drawing combs first themselves. So your frames of foundation may not receive a warm welcome. Because of this, don't add supers of foundation too rapidly during a weak nectar flow. You can get away with it during a strong flow. After your first season, when you'll have drawn combs to spare, you can be more creative in supering.

Keep an eye on the queen's brood pattern to make sure it's uniform and contains numerous eggs.

Once the temperature climbs above 65°F, you can safely leave the hive open and remove frames to examine them.

In May, start taking measures to prevent mites and continue your management program throughout the summer.

Continue to check for queen cells (a sign of overcrowding).

Late Summer/Fall

To encourage the bees to consolidate their stores, start adding supers between the brood nest and the partially filled supers (undersupering) in July.

Continue to check for queen cells and remove sealed supers as needed. Freeze the honey to eliminate wax moths.

Continue to monitor the colony for disease and mites. Treat your bees for mites in late summer, at the latest in early fall. Feed them some syrup medicated with Fumidil-B® for nosema.

To prevent robbing, don't work the bees unless you have to.

Remove all surplus honey and any empty or partially filled supers. Remove queen excluder(s). Make sure the colony will be able to overwinter on fully drawn combs. Reduce the hive entrance and, if you area has cold winters, insulate the hive.

Apply the tactics you learned in Chapter 5 to keep the mice out.

Add a top entrance to the hive for ventilation. Bore a 5/8 inch hole through the top hive body near the hand hold, or insert a small stick or stone under the front edge of the inner cover. In the latter case, push the outer cover forward so the bees can fly through the gap in warmer weather. Both the top and bottom entrances should be in the front of the hive; otherwise there'll be a draft. Make sure the bottom board slopes for drainage.

If your colony has stored at least 60 lbs of honey, they need a space of two to three hive bodies, as well as multiple frames of pollen in the brood nest. The whole hive should weigh about 125 lbs.

You need to distribute the honey stores so that the cluster will have at least 40-45 lbs of honey above it during the winter. The cluster itself will need some empty comb, as bees can't cluster on capped honey. Replace a couple of the frames in center of the upper hive body with partially filled ones from the lower body. Add the removed frames to the lower body.

If your colony hasn't been able to store enough honey, feed the bees heavy syrup during September/October until they have at least nine full-depth frames of honey. Let the syrup cool first before giving it to them. Each gallon of syrup will increase your bees' reserves by seven pounds. As mentioned above, the syrup can contain Fumidil-B®.

Ideally, you should requeen each colony once a year, at most every two years. This stabilizes production, controls swarming, and combats disease. To find out how to requeen a hive, go to the "Further Reading" section at the end of the book.

Once you've removed all the honey supers from your apiary and stored them under favorable conditions (such as freezing), you need to process them. In Chapter 3, we recommended using thin foundations in shallow honey supers for producing cut-comb honey.

Before cutting the comb out of the frame, decide which side looks the most appealing. Then use a sharp knife or comb cutter to remove the comb. Don't use the edges of the comb, and remove any uncapped cells. When you're satisfied with the pieces, place them on a wire rack over a pan to drain overnight at a warm temperature. Store them in plastic containers in a freezer.

In subsequent seasons, you can purchase better equipment for extracting and processing honey for sale.

The Sweet Taste of Success

Congratulations on completing this elementary course in beekeeping! We hope the preceding pages have been informative and encouraging. While other so-called "basic" beekeeping books present advanced concepts alongside fundamental ones, we sought to provide a true beginner's guide, including only those concepts you should know for your first season as a beekeeper.

After mastering the techniques described in this book, you can move on to more complex beekeeping. You'll be able to expand and strengthen your apiary through equalizing, splitting, and requeening colonies. Moreover, you can explore new methods of honey processing and perhaps sell the honey under your own label.

If anything, beekeeping requires patience, a willingness to refrain from merely reacting to problems and to instead put in the work of preparation and prevention. It also involves a trial-and-error style of improvement and problem-solving.

Though supervising and caring for thousands of little buzzing bees can be challenging, this very aspect can be the most rewarding. In what other activity can you observe and participate in a fully functioning, model society of insects? The beauty of a honeybee colony may take your breath away sometimes! Consider its simultaneous resilience and vulnerability, its industriousness, hierarchy, and communication, not to mention the social dramas continually playing out within and between colonies, and the constant battle with predators.

Beekeeping is animal husbandry at its finest. You should feel a sense of pride every time you put on your hat and veil!

Further Reading

Comprehensive:

Bonney, R.E. (1993). *Beekeeping: A Practical Guide.*

Bonney, R.E. (1991). *Hive Management: A Seasonal Guide for Beekeepers (Storey's Down-To-Earth Guides).*

Collison, C. H., Frazier M., Caron, D., Harmon, A., & VanEnglesdorp D. (2004). *Beekeeping Basics.* Mid-Atlantic Apiculture Research and Extension Consortium.

Cramp, D. (2012). *The Complete Step-by-Step Book of Beekeeping.*

Flottum, K. (2010). *The Backyard Beekeeper - Revised and Updated: An Absolute Beginner's Guide to Keeping Bees in Your Yard and Garden.*

Sammataro, D. (2011). *The Beekeeper's Handbook.*

Wiscombe, D., Blackiston, H. (2011). *Beekeeping for Dummies.*

Requeening:

http://www.beekeeping.com/articles/fr/selection_%20queen_breedings.pdf

http://basicbeekeeping.blogspot.com/2007/12/lesson-19-requeening-hive.html

Choosing a breed:

http://www.diablobees.org/beeinfo.html#20

http://homesteadrevival.blogspot.com/2012/03/choosing-bee-breed.html

General:

http://www.backyardbeekeepers.com/facts.html

http://beehivejournal.blogspot.com/2010/05/drawn-comb-vs-new-foundation-during.html

http://www.glenn-apiaries.com/beginning_beekeeping_package_bees.html

http://www.beehacker.com/wp/?page_id=22

http://www.dummies.com/how-to/content/where-to-keep-your-hives-when-beekeeping.seriesId-249743.html

http://cookevillebeekeepers.com/06/inside-your-hives

ABOUT THE AUTHOR

Nicole is an adventurous outdoor gal who enjoys pursuing homesteading related hobbies. If it involves keeping bees, raising chickens, or organic gardening than you can expect to see her there. If she is not out chasing the chickens or writing, then she is probably chasing around her two boys or husband.

www.ingramcontent.com/pod-product-compliance
Lightning Source LLC
Chambersburg PA
CBHW071818170526
45167CB00003B/1355